Abel Hernández-Muñoz

Treasures in the rocks

Abel Hernández-Muñoz

Treasures in the rocks

My first Paleontology

ScienciaScripts

Imprint

Cover image: www.ingimage.com

This book is a translation from the original published under ISBN 978-620-2-11476-9.

Publisher:
Sciencia Scripts
is a trademark of
Dodo Books Indian Ocean Ltd. and OmniScriptum S.R.L publishing group

120 High Road, East Finchley, London, N2 9ED, United Kingdom
Str. Armeneasca 28/1, office 1, Chisinau MD-2012, Republic of Moldova, Europe
Printed at: see last page
ISBN: 978-620-5-68343-9

TREASURES IN THE ROCK

ABEL HERNÁNDEZ MUÑOZ

INDEX

WHAT IS PALEONTOLOGY?

Much is written for young people about living animals, but very little is written about prehistoric animals, that is to say, those that existed in past times and have become extinct; that is to say, about those beings that the common people have come to call antediluvian, believing in good faith that the universal deluge put an end to them all at a stroke. On the other hand, there are few books in Spanish on this subject and written especially for young people; and it is not because the subject is unattractive, on the contrary: I know from experience that there are few questions about the universe that fascinate and excite the layman as much as the knowledge of the animals that lived in the remote past. Wishing to fill, at least in part, this void, I have written the present little work; but in it should not be sought even the intention of presenting an elementary treatise on paleontology; my purpose has been none other than to try to awaken the interest of young people towards the past of the animal world, presenting them in the most pleasant way possible, the most remarkable and interesting aspects of those extinct faunas. Consistent with this purpose, I have tried to avoid all technicalities and not to involve the reader in complicated biological or geological problems, in the certainty that only in this way is it possible to divulge a science such as paleontology, of which the public, in general, has only a few vague ideas, and almost always mistaken ones. I will end these preliminary words by stating that paleontology is the science that deals with the study of fossils and therefore of prehistoric plants and animals.

A LITTLE BIT OF HISTORY

For a long time men of science had wondered how old the Earth might be. Some thought it was only a few million years old and had always been just as they knew it, with the same animals and the sea and the Earth occupying the same place. Then they noticed that the rocks were made up of layers of different colors and sometimes had shells, bones and the delicate outlines of something that looked like leaves. But they believed that these "fossils," as they called them, were simply vagaries of nature.However, about 150 years ago, some scientists began to realize that fossils actually represented plants and animals that had lived a very long time ago. They also suspected that the Earth was much older than they had previously imagined. They also understood that the transformations of the Earth could only be known by studying the rocks that provided the key. So they set out to gather all the clues.The scientists who analyzed the shells from the reefs of Lvme Regis discovered that these shells corresponded to shellfish from very remote times. When, at the age of twelve, Mary Anning had discovered those vertebrae, she had been the first to discover a complete skeleton of an ichthyosaur, a vertebrate or animal with a backbone, which had died millions of years ago.In other parts of the world fantastic discoveries were also being made and little by little the history of the Earth was being reconstructed, which, according to scientific men, occurred in this way:

- Foundational stage. In 1796, at the beginning of the 19th century, Cuvier (1769-1832) published his work Memoire sur les especes d'Elephants tant vivantes que fossiles, which marks one of the main milestones in Paleontology, since it provides irrefutable evidence in favor of extinctions for the first time. On the other hand, his work on comparative anatomy and functional morphology, makes Cuvier to be considered as the founder of Paleontology, by providing it with a series of basic principles for its research and, in turn, of Paleozoology. or Paleobotany. His contemporary Lamarck (1744-1829) was the first to develop an evolutionary theory; however, neither his arguments nor the evolutionary process itself was accepted by his contemporaries, and one of his main opponents was Cuvier himself, a staunch defender of the catastrophist theories.

Throughout the 19th century there was a great proliferation of important works in paleontology. Undoubtedly, the work of Charles Lyell and other great geologists of the time paved the way for Darwin to elaborate his theory of evolution. This marked the beginning of a new stage in paleontology. With the publication of On the origin of species by means of natural selection in 1858, a true revolution took place and the beginning of a new and flourishing era for Biological Sciences, as well as the divorce between Paleontology and the other Life Sciences. Although Darwin had based many of his conclusions on fossils, it was paleontologists and geologists who took the longest to admit his theory. At the end of the 19th century and the beginning of the 20th century, with the beginning and development of Genetics, the greatest disharmony occurred; while Paleontology focused on stratigraphic studies, integrating itself into the Geological Sciences, Biology ignored Paleontology, considering it a purely descriptive science.

- **Modern Stage:** Thanks to the joint efforts of some biologists and paleontologists there is a reunion between both sciences within the framework of the new synthetic theory. Simpson with his work Tempo and mode in evolution (1944),[10] will be the precursor of this reconciliation that initiates a new stage in modern paleontology and the development and consolidation of paleobiological studies.

If the sixteenth to eighteenth centuries were characterized by large systematic studies and the nineteenth and early twentieth centuries by their applications in biostratigraphy, it is very recently when an important turn in paleontological studies took place. It was probably triggered by the theory of plate tectonics, to which paleontological studies made an important contribution through their paleobiogeographical contributions. Another factor, perhaps even more important than the previous one, has been the rapprochement of paleontology to the sciences. Biology, from which it had distanced itself since the last century. Paleontology is currently nourished by new techniques (electron microscopy, X-rays, spectrometry, computer science) providing new and interesting data in various paleobiological aspects (Paleoecology, Taphonomy, Paleohistology, Paleobiochemistry...).) The studies of protists, pollen and fossil spores, widely developed since the second half of this century, have been a very important complement to classical paleontological studies, with contributions in the field of the origin of life, evolution, Taphonomy and Applied Paleontology, among others. Paleobiochemistry studies are currently experiencing a remarkable boom, opening a new field of research with great possibilities in various paleobiological aspects. In the field of evolution, the theory of punctuated equilibrium (Eldredge and Gould, 1972) has burst with force in recent years, criticizing the synthetic theory and creating a lively controversy.

THE HISTORY OF THE EARTH

The Earth is more than four billion years old. At first, it was a ball of gas that lasted millions of years. Gradually, as it cooled, it became a fiery sphere of sheared minerals. Then very slowly, the heavy minerals sank toward the center of that sphere, and the lighter ones remained on the surface. The granite formed blocks that would later become the continents between which deep basins remained.

Many years passed until finally the vapors emitted by the earth reached the highest and coldest layers of the atmosphere, which condensed into water. Then the rain began to fall. It rained so much that the enormous basins gradually filled up and the oceans were formed.Finally, after millions of years, the Earth had its continents and seas. But it was a strange and desolate Earth. On the bare rock of the Earth, there was not a bit of moss. In the ocean there was not even the most insignificant shell or sea plant. There were no singing birds or roaring animals. There was only the lapping of the waves against the bare rock without the slightest hint of life.It took another million years before at some time and somewhere in the water life was produced. No one knows how this happened, or what its first manifestations were like. However, over the centuries, many kinds of plants and animals evolved first in water and then on land. When man emerged, barely a million years ago, many different types of plants and animals had already managed to live for a period of time. Then they became extinct forever. We only know that they existed because of the traces they left us in the rocks. They are called prehistoric plants and animals because they lived long before history began to be written.

THE READING OF THE ROCKS

The rocks that tell us about prehistoric animals were formed in a very special way. The rain that fell on the earth's crust and the ice that cracked on it were detaching small pieces of rock and earth. As they were carried away by the currents, they were also wearing away the earth's crust. Soon, the rivers were filled with mineral particles and grains of rock that, carried by the current, reached the oceans and lakes.Over millions of years, these mineral particles, together with small shells of aquatic animals, formed layers at the bottom of lakes and oceans. The material that accumulates is called sediment. As layers accumulate on top of each other, the lower sediments harden to form sedimentary rocks.Usually, when plants or animals die, they rot and disappear. Sometimes, however, when a plant or animal dies, it is accidentally covered by a layer of mud or sand. Sometimes this layer prevents decomposition and the dead animal or plant remains intact for some time. Then, as thicker and thicker layers form, the tracks of the animal or plant are preserved as fossils in the sedimentary rocks.The word fossil comes from the Latin word meaning "unearthed". There are many kinds of fossils. Some are the original bones or shells that were preserved in the rock. In other cases, the shell or bone material was partially or completely replaced by minerals that leached out with water, and the fossil turns out to be a mineral model of the original. Or, on other occasions, the plant or animal disappears, leaving the mold or footprints. Fossils are all traces of a past life preserved in the earth's crust.

Many of these fossils originated in shallow sedimentary layers of water. Later, as the earth's crust contracted and folded, these layers rose above the water level forming irregular folds and twisted like mountains. Immediately the wind, rain and other elements began to wear away these layers of rock until the fossils were exposed.Scientists can calculate almost exactly the age of a rock layer. Of course, if they know the age of the rock they can also determine the age of the fossil animals and plants it contained.Through the study of rocks and their fossils, men of science have been able to divide the history of the Earth into various periods by separating them by the intervals at which the mountains were formed and by the different kinds of fossils. They can determine with precision the period in which each fossil lived.Of course, no man of our time has ever seen these prehistoric animals. The representations we make of them are made following the traces they left behind. However, the sources we possess are many. There are complete skeletons of animals that give us an exact idea of their dimensions and the distribution of their bones. There are also footprints of some animals that show us what their skin looked like. Therefore, the representations that are made today are undoubtedly very similar to what the animals were really like long ago.

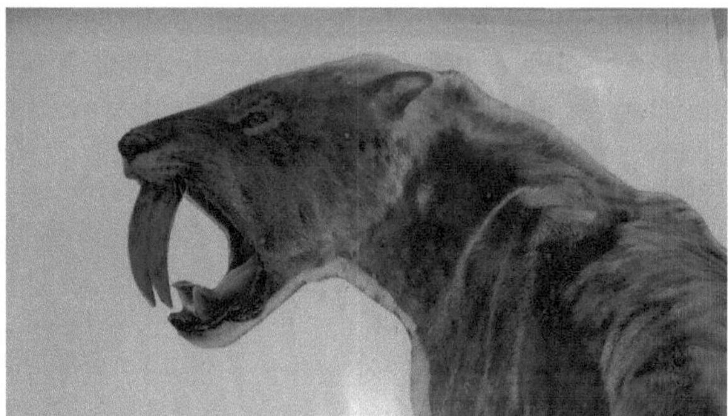

LIFE AND ITS ORIGIN

On the ground, in the water or in the air there are millions of living things. Look around you! Can you distinguish what is alive? What is the difference between a living thing and a non-living thing? Humans, animals, plants and many other organisms that our eyes cannot see are alive. Why do we say they are alive?

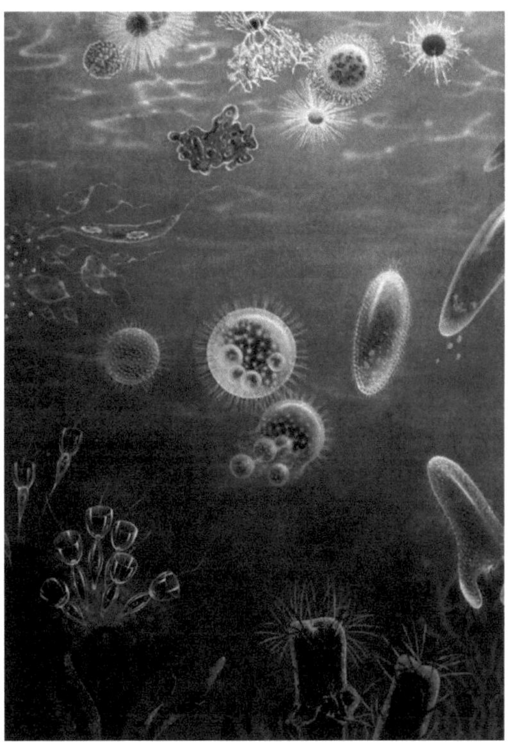

THE ORIGIN OF LIFE

Scientists think that about 3.8 billion years ago the first forms of life on Earth emerged. It is believed that the first living beings lived in the water of the oceans. They were very small organisms and different from the vast majority of those that now populate our world.

WHAT DOES IT MEAN TO BE ALIVE?

Have you ever thought about what you need to be alive? You need to breathe, feed yourself and eliminate certain substances. You need energy to move, jump or run. Your body requires energy to live. In addition, being alive is also about relating to the environment in which you live and responding to changes in the environment. For example, when it is hot, your body starts sweating to lower the temperature, and you take shelter in the shade. To be alive is also to grow and reproduce, that is, to have children. We call all these activities the processes of life, the **vital functions.** In short, being alive means being able to perform all these functions.The vital functions of living beings are: the **function of relating** to the environment in which they live, the **function of reproduction** and the **function of nutrition.** The function of nutrition allows us to obtain energy, develop and grow, and includes respiration, digestion, transport of substances necessary for life and excretion or elimination of wastes. Living beings are born, grow, reproduce and die.

LIVING AND NON-LIVING BEINGS

All organisms that are alive perform, albeit in different ways, all the processes of life. Humans, animals, plants and many organisms that our eyes cannot see have life.However, the stones, the air, the earth or the objects we manufacture have no life. They do not grow, they do not reproduce, they do not need energy, they do not respond to the things that happen in the environment where they live. They are not living beings.

THE CELL AND LIVING BEINGS

Another characteristic of living things is that they are all made up of small units called **cells.** Some living things, such as bacteria, are composed of a single cell; they are **unicellular** organisms. Others, such as plants and animals, are made up of more than one cell, even millions of cells; they are **multicellular** organisms. Cells are alive because life processes also take place in them.

DIFFERENT LIVING BEINGS

Although all living beings perform common vital functions, they are not all the same. Each type of living being has different characteristics and a different way of carrying out vital functions. For example, plants feed on water, the sun and substances from the earth; however, animals feed on plants or other animals.The different types of living things are classified and grouped according to the characteristics they share. The largest groups of living things are called kingdoms. You know some of them, such as the animal and plant kingdoms. The other kingdoms are the fungi, protists, and prokaryotes.

THE FIRST SETTLERS OF THE EARTH

While the fish were conquering the oceans, something even more impressive was happening. A few pioneers emerged from the water and ventured onto the bare land. A few green seedlings had been growing on the muddy shores for some time. During the Age of Fishes, the first forests of huge fern-like plants appeared.The animals were also looking for ways to get ashore. Even before fish were abundant, small scorpions, four centimeters long, were already scurrying along the shores. Apparently, they were the first animals on Earth to breathe air. Later, snails appeared. But it was a long time before vertebrates appeared. For a vertebrate animal to live on the surface of the earth it would need strong legs to lift it off the ground and allow it to walk. Its body would need to have a bone structure to give it shape. It would also need lungs to breathe air.Fish were the vertebrates of the time, but they lacked legs and most breathed through gills, not lungs. The gills are moist folds of muscle tissue or membrane, with slits running from the fish's throat to the water. The fish swallows the water, which then passes through the gills. The gills absorb oxygen dissolved in the water.

THE NEW TYPE OF ANIMAL

As the centuries passed, vertebrate animals began to walk on Earth. In rocks from 200 million years ago, the first footprints have been discovered. The prints show that the hind feet had five toes and the front feet had four toes. It is believed that these terrestrial vertebrates evolved as follows:During part of the Age of Fish, the Earth was flat and wet, with many inland bays and lakes. But then the Earth entered a period of retreat and mountains formed as the land partly retreated. Some lakes and bays gradually emptied until they were dry, because there was very little rain at that time. Some fish managed to swim to deeper places. Many in stagnant, swampy water died. But some found a way to adapt to this drier environment.To this category belonged the crosopterygians, fish with lobe fins. They had four short fins strengthened by a bone structure. They had a backbone and a kind of lung that allowed them to breathe air.Gradually the lakes continued to dry up and the fish were dying. But some of the crosopterygians crawled painfully inland and centimeter by centimeter, they crawled along the land with their short fins. The strongest made it to a lake where they could rest at least for a while.This went on for a long time. The stronger lobe-finned fish, better adapted to this new dry climate, managed to survive. Gradually some were transformed. They began to have legs and lungs bones bodies more and more suitable for living on Earth. Finally, after millions of years, some fish had changed so much that they formed another type of animal called an amphibian. This name comes from the Greek which means one who "lives a double life". Amphibians were able to live on land and in the water.Of course the crosopterygians didn't transform because they wanted to. It's just that they could not avoid change. Among the animals of the same species, some are born with small differences with respect to the others. Sometimes the changes are not advantageous for the animals and so they die. But at other times, the opposite happens. Then the animal not only survives, but adapts to the environment very well. Possibly the offspring of this animal will also undergo small changes that benefit them in the same way. Their descendants may change even more and finally, after millions of years, the sum of these small changes will result in a completely different species of animal. This is what happened to the crosopterygians that became amphibians.

SWIMMING AND FLYING

Now the reptiles evolved into different kinds and sizes. Some returned to the water. The ichthyosaurs evolved into streamlined fish-like animals. It was the bones of one of these animals that Mary Anning discovered in the rocks of Lyme Regi. They were animals from one and a half to three meters in length, although some varieties reached up to nine meters.The ichthyosaur moved rapidly through the water with undulating movements of the body and tail, as fish do. Its legs gradually decreased in size until they became short fins, which only served as a rudder. It had a long, strong snout and curved teeth that helped it retain the fish it fed on. Its enormous eyes probably located the prey from far away and could propel itself quickly after it thanks to its strong tail.Very different from the agile ichthyosaurs were the clumsy plesiosaurs. Another type of aquatic reptile with a barge-like appearance, it probably swam slowly through the water with its four short oar-like legs.Some of the plesiosaurs reached about 12 meters in length. Much of this length was often due to their long, flexible necks. A plesiosaur was discovered whose head was half a meter long, the body about three meters, the tail two meters and the neck seven meters. A man of science once said that the plesiosaur must have looked like a snake emerging from the body of a turtle. Often the neck comprised more than half of the total length of the plesiosaur.

That long neck was very useful. The plesiosaur was a slow-moving animal that splashed majestically through the water. But as it splashed about, it could suddenly stick its head in the water and devour a fish swimming a good distance away.Another type of reptile, the pterosaur, took to the air. They came in many sizes, some the size of a sparrow and others measuring six meters from wing tip to wing tip.Pterosaurs did not have wing feathers like birds. The wings of the pterosaur were made of skin attached, on the one hand, to the fourth toe of the animal's front legs and, on the other hand, to its hind legs. The front claws were short and curved and allowed it to walk or hang upside down from tree branches or boulders.The teeth of some of the earliest pterosaurs were needle-like. The tail was long and had a kind of paddle at the tip. Over time they changed a lot and later, pterosaurs had large beaks without teeth and a very short tail. They probably lived in trees or on reefs and fed on flowers, fruits and insects.

Reptiles that flew through the air and glided over the water must have been a common sight in those times. But even more important than these creatures were the first land reptiles. Some of these were the ancestors of the mammals that appeared much later on Earth. At that time, however, only one type of land reptile was of special importance because it came to dominate the Earth for 125 million years longer than any other animal that ever lived.125 times more than man! These reptiles were the dinosaurs. The period in which they lived began about 200 million years ago and is called the "Age of Reptiles". When men of science began to discover in the rocks the strange fossils of the "Age of Reptiles". they said, "There is no name for these animals." How shall we name them One came up with the name "dinosaur" two Greek words meaning "terrible lizard".

Since then, when a new dinosaur is discovered, its discoverer gives it a special name, generally using Greek or Latin words as is customary for classifying plants and animals. Although these names may seem complicated to us, they actually make things easier since they determine, once and for all, the name of the dinosaur in any language, instead of having one name in English, another in German and so on.

BIRDS? NONSENSE!

More than a hundred years ago in the Connecticut River Valley of New England, farmers plowing the land and quarrymen quarrying the stone often found footprints that had been imprinted when that rock was nothing but soft mud. The footprints corresponded to animals walking in a straight line on two legs that resembled bird claws with three fingers. Some were as small as the footprints of a chicken, others were huge.

"So many birds! "How is it that there are so many sizes? In what era did they live?

No one could explain it until a man of science came along who knew a lot about rocks.

"Birds! " he grumbled. "Impossible! Those rocks are too old. There were no birds yet when those footprints were formed."

He examined them closely. From time to time another mark appeared behind the footprints, suggesting that the animal had dragged a long tail. And along with the three-toed footprints there was another four-toed footprint that undoubtedly belonged to a reptile.

"These are not bird tracks," the man of science said. "They are dinosaur tracks like those that have been discovered elsewhere in the world."

More men of science began to investigate in that place. They finally succeeded in discovering some dinosaur bones, but they never found bird bones in such ancient rocks. In fact the Connecticut swamps, turned to rock, were full of dinosaur footprints, which I saw in the age of reptiles.

THE PRIMITIVE SAVAGES

Primitive dinosaurs were light and fragile animals, some about the size of a chicken. They did not have legs on their sides as primitive amphibians and reptiles did. Dinosaurs had legs under their bodies. This means that dinosaurs could run on their hind legs by placing one leg in front of the other.Given their light constitution, these animals, with their bird-like legs, could run like arrows, swinging with their long, stiffly raised tails. Apparently they could not jump, but only walk or run. Their front legs were much shorter than their hind legs, and although they never used them for walking, they may have been very useful for holding food and for fighting. These dinosaurs had the sharp teeth characteristic of carnivores. It is very likely that small reptiles ate the eggs of other dinosaurs.The smallest known dinosaur is the compsagnato-found in limestone rocks in Germany. This little animal measures less than a meter from head to tip of tail. Some of the tracks discovered in the Connecticut river valley appear to be of even smaller dinosaurs. Although none of their bones have been found. These tiny reptiles look like toys next to the enormous allosaurus that was the terror of its time. More than ten meters long, this roaming colossus appeared standing on its powerful hind legs, opening its hungry snout wide and swinging its body with the sway of its rigid tail. With strong hooked claws and knife-sharp teeth, it tore the flesh of the reptiles on which it fed.

However, the largest and most fearsome of all terrestrial carnivores was the tyrannosaur that lived at the end of the Age of Reptiles. This monster was fourteen meters long and reached almost six meters in height. Two powerful hind legs held it upright in an almost vertical position. The enormous head of the tyrannosaur measured more than a meter in length and had jaws like steel, equipped with sharp teeth that measured up to fifteen centimeters. Its strong legs and hind claws held firm its formidable stance and, although its forelimbs were so short that they did not even reach its snout, they also had tremendous claws to tear the enemy to shreds.

ARMORED ANIMALS

Some dinosaurs could not flee to the lagoons to defend themselves from the fierce carnivorous reptiles. Several herbivorous dinosaurs had heavy armor to defend themselves.The stegosaurus was an animal about six meters long that moved slowly on its four legs. This reptile had a small head that was lost in its neck. When the stegosaurus was planned, the plans apparently did not include a place for the brains. This animal must have had almost no brain at all.However, it had armor along its back, made of triangular-shaped bony plates. On its tail it had four menacing beaks half a meter long. The huge hind legs raised its body much higher at the rear than at the front, supported by much shorter limbs. That's why it looked arched. When any enemy attacked it, it was defended by its arched spine with its strong bone plates.The tremendous turtle-shaped ankylosaur had better defense. Its body was enclosed between closely packed bony plates. Rows of sharp beaks protected its sides, and at the tip of its tail was a large bone club that it used as a bludgeon. Whenever a problem arose, the ankylosaur would stand still. A bludgeon was enough to scare the enemy away.

Other dinosaurs had horns and large bone shells. One of these was the triceratops, about six meters long, which, unlike many others, had a huge head comprising almost a third of its length. The snout was pointed like a beak with a small horn on top and two longer horns above the eyes. Behind the eyes, a huge carapace of bone extended backwards and protected the neck. This imposing animal must have fought a lot, for its bones almost always bore traces of scars.The last reptiles to appear were those with horns. Then, in a short time about 7 million years ago, all the dinosaurs became extinct.No one will ever know exactly why the dinosaurs disappeared so completely. Perhaps it was due to their enormous size and low intelligence. It is also possible that many factors were involved. But what is certain is that by the end of the Age of Reptiles new mountain uplifts had occurred. The bottom of the inland seas was rising above the water level. The seas retreated further as the land was rising. The warm swamps that the dinosaurs loved so much eventually dried up.As the highlands appeared, the days and nights became colder. The plants on which some dinosaurs lived gave way to others that could live in colder climates. Little by little the dinosaurs lost not only their food, but also the warm climate to which they were accustomed. Thus they disappeared. And along with them, of course, the carnivorous dinosaurs that they were used to feed also became extinct.

DESCENDANTS OF REPTILES

Only a few reptiles have managed to survive, such as snakes, lizards, turtles, alligators, crocodiles and the New Zealand tuatara. However, the descendants of those remote beings, including birds, surround us everywhere.

In 1861 stonemasons in Bavaria found, in limestone rock, a strange Fossil in a layer in which there were usually only the remains of small reptiles. It was the cast of a feather - a single feather! Here was something new! No one dared to say that it was a feather because so far no one had assumed that birds had lived in the reptilian age. However, a month later another fossil was found that looked like a small reptile with wings and a feathered tail. The scientists decided that it was a remote bird and called it archaeopteryx, which means ancient feather.

MAMMALS EMERGE

When the dinosaurs disappeared, another class of animals took over the Earth: mammals.Mammals represented as great an advance over reptiles as reptiles did over amphibians. Reptiles were cold-blooded, meaning that their body temperature was more or less the same as the surrounding environment. If it was freezing, they froze, if it was too hot they overheated. In both cases they died. Mammals, on the other hand, were warm-blooded. Their bodies always remained at the same temperature, so they could survive in cold or hot weather.Reptiles generally hatched from eggs laid in the ground and had to fend for themselves from the moment they hatched. But the eggs often broke or were eaten by other, larger animals. Mammals did much better. Most of them carried their young inside their bodies for some time where they were free from danger. When the young were born, the mother would feed the young with milk from her own body. She cared for her young until they could fend for themselves.Mammals also had other advantages. The r had hair that protected their bodies, while reptiles had only smooth or scaly skin. Mammals also had larger brains.In reality, there was nothing new about mammals. They had already begun to exist millions of years before, since their ancestors were the primitive reptiles, and they had been roaming the Earth almost since the time of the dinosaurs. But the first mammals were small, most of them the size of a mouse. The small mammals could not stand up to the monstrous dinosaurs during the entire era when they lit up the Earth. These little animals hid as much as possible. Many of them probably scurried about on the ground, hidden by the shadows of the night, and others, no doubt, hid in the trees. They probably fed on insects, nuts, juicy bark and the fruit that was already on some trees.

Over millions of years, the small animals were constantly progressing. Slowly, their brains memorized and their bodies became more and more abandoned to the environment. When the dinosaurs became completely extinct, the mammals were ready. They left their hiding places and spread out over the entire surface of the Earth. This was about 75 million years ago.

THE SMALL PRINCIPLES

Rhinos also kept changing. The swift rhinos of the ancient forests were companions of some of the earliest horses, looking very much like them. These small rhinos were very slender and lacked horns.Gradually, however, the rhinoceros took on various forms. In general, it became small and stubby. Some had no horns and others had only one. The best known rhinoceros lived about 20 million years after the extinction of the light rhinoceroses. It was the size of an adult pig. It had a barrel body with legs that looked as if they had been sawed off and two short horns on each side of its snout. Surely large herds of these little animals roamed America some 30 million years ago, as the bones of about 16,000 were found in one place in Nebraska. The place where they came down to drink was transformed into quicksand that buried them one by one, as they came down to drink water.Camels were also American animals that came into existence more than 40 million years ago. In those days camels were small, delicate animals, about the size of a rabbit.Later they grew larger and changed in appearance. Many were sheep-sized and looked much like today's South American llamas. Some had legs suitable for walking in dry sands.Other camels had long necks and legs like giraffes. They lived in forests and ate leaves from trees. But all camels lived in North America for millions of years. They moved to other continents in very recent times.

All major groups of mammals appear to have started out as small animals, some of which increased in size over millions of years. The remote ancestors of the tuskless, trunkless elephants followed in those same footsteps. Twenty million years after living in Egypt, some of their ancestors, the mastodons, arrived in North America from Asia. At that time there was a strip of land in what is now the Bering Strait. These mastodons looked like long-haired elephants: They were two to three meters tall. They were about the size of today's Indian elephants, but were heavier in build. They had two tusks sometimes reaching almost three meters in length. Some mastodons lived on the plains. Other large herds roamed the forests eating acorns. Their bones are often found in swamps that have dried up. One skeleton was found so complete that part of its thick, woolly fur still hung from it. Woodland mastodons first arrived in North America about 20 million years ago and their descendants must have lived until about 620 million years ago.

NORTHERN ICE

The Earth was getting colder and colder until a million years ago, when huge glaciers began to spread over some regions of America, Europe and Asia, some parts of the southern hemisphere.Four times in a period of seventy thousand years, glaciers, huge masses of ice, invaded part of the northern continents. They remained there for a long time until they finally melted. Even at a great distance from them the Earth was frozen and this affected the animals.From the beginning of the Ice Age, animals began to migrate. Camels began to move away from North America. Some, like llamas and their relatives, fled to South America. Others went to Asia and Africa. North American horses spread to other regions of the world. Elephants called rnamuts came to North America from Asia, while megatheres and glyptodonts came from South America.Megatheres were fat, clumsy animals, some up to six meters long. They had claws that were like large hooks and bones in their feet so close together that they had to walk by leaning on the knuckles of their front feet on the outer sides of their hind feet. Some varieties of megatheres had bones like plates embedded in the skin.When the megatherium was hungry, it would stand up on its hind legs, roll its tongue in a bundle of tree leaves and put it in its mouth.

The descendants of these animals lived until so recent times that we know that prehistoric man knew them. In one South American site human bones were found along with the reddish brown bones of a megatherium. Some of the bones of the megatherium indicated that man had broken them to remove the marrow.The South American fiddurids were distant relatives of today's armadillos. They lived in bone shells that measured about three meters long by one meter high. The tail of jointed bone rings often had spikes at the tip. When attacked by other animals, the dedichurans They simply hid under their armor and lashed their enemies with their hard and heavy tails.There were several types of mammoths at that time, majestic beasts somewhat similar to the elephant, with huge twisted tusks. The largest of all was the imperial mammoth that lived in the glacial period and often reached a height of nine meters at the shoulder. As this animal aged, its immense tusks became twisted and completely useless for digging or protection from enemies. At one time, the mammoth existed in Siberia, Europe and North America.The fiercest animal of this time was the smilodon, which was about the size of a lion. Its fangs measured up to twenty-five centimeters in length and had the cutting power of a dagger. In addition, the edge of its fangs had saw-like teeth. Because its fangs were extra long, the smilodont's jaw did not function in the usual way. It had a special constitution that allowed it to open its jaws at right angles so that its sharp fangs could spring into action.

IN THEIR ENVIRONMENT AMONG THE GLACIERS

Megatheres, deciduids, mammoths, and smilodonts lived far to the south away from the icy world of the glaciers. At that time there were two beasts that could withstand the icy blasts that blew from the great expanses of ice. These were the woolly rhinoceros and the woolly mammoth that had a thick fur with a kind of woolly lining. We know what these two animals looked like because we have detailed paintings made by the prehistoric men who saw them. Some of them, who lived in France and Spain, mixed brown and yellow earth to prepare paintings. On the walls of the caves where they took refuge, they painted majestic processions of these animals. Those who admire these paintings today are amazed by their beauty and accuracy. Those prehistoric men were artists who drew the animals they saw frequently.A few years ago, the intact body of a woolly mammoth was discovered in the frozen ground of Siberia. The animal had fallen into a crevice in the ice many centuries before and had remained there ever since, preserved intact. as in an open-air freezer. Those who dug it up observed that it had a broken leg and rib, and that a blood vessel had burst while it was struggling to get out of the crack. It retained in its mouth some of the grass it was eating at the time of its fall. On examining the mammoth's stomach, the scientists found that it ate grasses and leaves that still grow in Siberia. For years, Siberians collected the ancient tusks of this animal to sell to ivory merchants. They say that more than 40,000 mammoths have been found.

FROM SOUTH TO NORTH AND NORTH TO SOUTH

As the ice spread southward, the animals overtook it, fleeing to warmer places. As the ice retreated, hordes of animals turned northward, following it as it resurfaced. Four times this happened. Horses, mastodons, megatheres, camels, tapirs, elk, deer, sheep, goats, oxen, giant beavers two and a half meters long, huge bison whose horns measured almost two meters from tip to tip, a host of smaller animals, all moved with the glaciers. Then followed the fierce hunters: bears, wolves, lions and other felines that never lost sight of them.But at the end of the Ice Age, many of these animals disappeared forever for no known reason. When the animals made a mass migration southward to escape to the glaciers, food was often scarce. This may have been one reason why some of them eventually became extinct. Perhaps the cooling of the climate was another reason. With the cold, some plants that were food for certain animals disappeared. At that time, man also appeared and, being the enemy of the animals, he may also have been the cause of the disappearance of some of them.Whatever the reasons, soon after the glaciers receded, times changed. Man was taking over more and more of the land, and animals were getting closer and closer to what they are today.

NEW DISCOVERIES TO BE MADE

Perhaps our knowledge about prehistoric animals is just the beginning. As more fossils are uncovered, we may learn new and exciting things about these remote beings. Scientists are always looking for new clues. Sometimes they stumble upon a new footprint, as happened in the case of the coelacanthid fish.One hot day in 1938, fishermen casting their nets in the sea off the South African coast pulled up a strange fish that measured five feet tall and had strong fins and powerful jaws. It was unlike anything they had ever seen before. The more they observed it, the more they became convinced that it was an important fish.Upon arrival at a small South African port, they rushed to the local museum to inform the curator of their find. She didn't know what the fish was either, but assumed it should be preserved for the specialists to see. As it was very hot, the fish began to rot and the museum attendant stuffed it as best she could.Finally, when the scientists observed the fish, they were astonished. This fish was a coelacanthid, a true living fossil, since it was supposed to have been extinct for 75 million years. It belonged to the family of ancient crosopterygians that first emerged from the water to begin the long procession of land animals.The coelacanthid had not transformed into an amphibian like the crosopterygian. However, if men of science could examine a coelacanthid well, they could conclude many things about the very ancient crosopterygian and perhaps about other prehistoric fishes. But this coelacanthid was mounted its valuable internal organs no longer existed. The scientists could hardly contain their tears. One of them, Professor J. L. B. Smith, decided, however, that where there had been one coelacanth there must be more. He made investigations for fourteen years, traveling thousands of miles up and down the African coast, publishing pamphlets describing the fish and offering a reward to anyone who found it.One day in 1952 a fisherman in the Comoros Islands caught another strange fish near the coast. The next morning when he was taking it to the market, a man pointing at it excitedly shouted: "Lots of money. from the professor's pamphlets recalled the prize on offer. The two men went together to see a friend of Professor Smith's and showed him the fish.That same day, when the professor was returning from a scientific expedition, he received a radio message telling him what had happened. He quickly left in a military plane for the small island where, in a sailboat, his friend was waiting for him with the fish.There on the deck was the fish, wrapped in absorbent cotton. The moment his friend unwrapped it, tears came to the professor's eyes. Fourteen years of searching had finally come to a successful end. Even though the fish was not in very good condition, by studying the coelacanth the scientists hoped to learn a lot about crosopterygians as well as other prehistoric animals.Who knows what other fossil treasures lie in the rocks! Recently in the state of Oregon, a seventy-year-old retired mail carrier had a hobby of collecting minerals and leaf fossils as a hobby. While removing a rock he discovered a strange tusk. Around him. Embedded in the rock were more bones. Experts have identified the letter carrier's find as 105 fossil remains of a mastodon from 10 million years ago - the only fossil of its kind! Prior to this, only a few teeth and a small piece of the jaw of this type of mastodon were known.The story of prehistoric animals never ends. Every year new chapters are written. And in some parts others still unknown are waiting for their turn.

THE EVOLUTION OF THE LIFE

All the species that populate our planet, animals, plants, fungi or bacteria, are slowly transforming. This phenomenon is known as evolution of species. For centuries, people believed that the world and living beings had been created by one or more divinities. However, at the end of the 19th century, a scientist named Charles Darwin developed a theory known as 'evolution by natural selection', which explained this process without the need for a creator. Darwin published his theory in 1859, in a famous book entitled The Origin of Species by Means of Natural Selection. Evolution has been one of the most important concepts in the history of science.

FOSSILS, EVIDENCE OF EVOLUTION

When **fossils,** which are the remains of living beings from the past preserved in rocks, are studied, it is observed that the most recent fossils are more similar to current species than those of older species. Thus, fossils show a progressive transformation of **species.** Thanks to them, it has been possible to reconstruct, for example, the long history of horses: it took 50 million years for a small herbivore the size of a fox, with five toes, to gradually transform into a large herbivore with long legs ending in a single hoof.

WHAT IS A SPECIES?

A **species** is a group of individuals that look alike, that share the same territory (which can be the size of a small region or an entire continent) and, above all, that can reproduce. But not all individuals of the same species are completely identical: they do not all have the same size, the same behavior or the same color. This diversity is due to small differences in their genetic material.

WHY AND HOW DO SPECIES EVOLVE?

When a cell divides in two, it makes two copies of its **genetic material** so that each of the two daughter cells has its own. In this copying process, an error can sometimes occur: one of the genes turns out to be modified and is not expressed in the same way; this is what is known as a **mutation**. If a mutation occurs in a sex cell, sperm or ovum, this error is transmitted in the offspring of the individual in which it has appeared. In this case, the mutation is said to be hereditary.Such errors in the genetic material often have dire consequences. But it can also happen that the individual carrying the mutation can live normally. In this case, it has a peculiarity that distinguishes it from the other members of its species. For this reason, all individuals of a species are not absolutely identical. In a few cases, this peculiarity (a different color, a slightly longer leg, better resistance to cold, etc.) gives it an advantage over its congeners. This novelty gives it a better chance of survival and, therefore, of reproduction. Since it will have numerous offspring, this new trait will gradually spread within the species to which it belongs. With the passage of time, new mutations will appear in some other individual of the species in question. The positive ones will endure and spread; the negative ones will disappear with the unfortunate specimens that possess them. Almost imperceptibly, the species transforms and develops new traits: it evolves.

THE IMPORTANCE OF THE ENVIRONMENT

In principle, no mutation is positive or negative, but is determined by the environment in which the species lives. Individuals with advantageous characteristics have more opportunities to live better and longer than others. Therefore, the environment tends to select individuals carrying favorable mutations and to eliminate those lacking them. This is known as **natural selection.** In this way, the species transforms and adapts more and more to the environment.

THE FORMATION OF NEW SPECIES

Certain mutations have more impact on some species than on others: individuals carrying them can only reproduce among themselves. Consequently, they are isolated from other members of the species. This happens, for example, when a species occupies a vast territory in which there are geographical barriers (mountain ranges, seas, etc.) that isolate some groups from others. The small isolated population will tend to develop characteristics different from those of the rest of the species, depending on the particularities of their environment. If this isolation is prolonged, there is a good chance that, with the passage of time, these individuals will not be able to reproduce with the other members of the species, even though, in the meantime, the geographical barrier has disappeared. In this case, a new species appears, different from the previous one.

THE CLASSIFICATION OF LIVING BEINGS

Thousands of different animal and plant species live in the world. Have you ever thought about the criteria used to classify them? Sometimes it is difficult to determine whether a living being is an animal or a plant. This is the case of the anemone; surely you have seen one on the seashore. At first glance, it looks like a flower with colorful petals; but it is not a plant, but an animal. One of the main tasks of biology is to classify living things into groups.

WHY IS A SCIENTIFIC CLASSIFICATION NEEDED?

The science that studies living things is called **biology.** One of the main tasks of biologists is to classify living things, i.e., to place them in a **group**. For example, a biologist studying an unknown living thing in the form of a worm must first find out what kind of animal it is: is it a caterpillar, a true worm, or a small snake? Scientists gather into groups those animals that have common characteristics.

SCIENTIFIC NAMES

Animals and plants have common names, but these names change from one language to another. For example, the animal known as elk in Spanish is called elk in English, élan in French, and elch in German. To avoid confusion, biologists have created **scientific names,** which are the same in all parts of the world. Thus, the scientific name for moose is Alces alces, and this can be used in any country by people speaking different languages. Sometimes the same common name can also be used to designate two different species. Such is the case, in Spanish, of the word 'langosta', which can refer to both a crustacean and an insect. As you can see, scientific names, which are the same in all countries, can avoid many mistakes and misunderstandings.

THE SPECIES

The **species** is the basic unit of classification of living beings; but how do you define a species? The brown bear and the polar bear, for example, are different species. They have many similarities, but also great differences: they do not have the same color; the polar bear is somewhat larger than the brown bear, and, in addition, the polar bear feeds on fish and seals, while the brown bear eats roots, fruits, insects and small mammals.For scientists, a species is defined by two characteristics: the first is that it groups individuals with very similar forms; the second is that these individuals can reproduce and have fertile offspring. A species can be divided into **subspecies. Within the** same species, animal or plant, sometimes there are groups of individuals that present differences, but that can have offspring: these are the subspecies, which are usually called breeds, in the case of domestic animals, and varieties, in the case of plants. Such is the case, for example, of domestic dogs: a St. Bernard, a poodle or a Dobermann are very different in character, shape and size; but they do not form different species: they are different breeds of the same species.

HOW DO YOU NAME A SPECIES?

The scientific name of an animal or plant is always written in italics and consists of two words, as established by the **binomial nomenclature** created in the 18th century by **Carl von Linnaeus.** The first, written with a capital letter, designates the genus; the second, in lower case, designates the species, generally citing one of its characteristics, such as color, size or region of origin. These two words are written in **Latin.** The human species, for example, has the scientific name Homo sapiens, from the Latin words homo, which means 'man', and sapiens, which means 'who knows'. Sometimes, the scientific name does not resemble the common name in English. This is the case of the black rhinoceros, scientifically known as Diceros bicornis (bicornis means 'having two horns'), and the peach tree, Prunus persica, which originated in Persia, now Iran. When a researcher discovers a new animal or plant species, he must give it a name following this model, which will be its official and international denomination.

LARGER GROUPS

Close species, which have a number of common characteristics, are classified into groups called **genera**. In species of the same genus, the first word of the scientific name is the same, while the second word is different. The tiger (Panthera tigris), the leopard (Panthera pardus) and the lion (Panthera leo), for example, are part of the same genus. The most similar genera are gathered in larger groups, which are called **families.** Thus, all felines form the Felidae family. The families are, in turn, included in the next group, the **order**. Felids, for example, belong to the order of Carnivores. The next group is the **class**. The animals mentioned above belong to the class Mammals, which includes many other animals, such as mice and gorillas. In total, there are more than 4,600 species of mammals. The last group is the **phylum,** which is composed of similar classes. Mammals, birds, reptiles, amphibians and fish belong to the phylum Chordates.

THE KINGDOMS OF LIFE

Finally, several phyla make up the largest groups of living things, known as **kingdoms.** The two main kingdoms are the **Animal kingdom** and the **Plant kingdom**. The Plant kingdom is also divided into phyla, classes, orders, families, genera and species.

Another of the kingdoms is the **Fungi,** which includes mushrooms, molds and yeasts. The difference between animals, plants and fungi is their source of energy. Animals obtain energy from the food they eat; plants, from sunlight; fungi, from other living things, breaking them down or digesting them on the outside and then absorbing them in small pieces.

Another of the kingdoms is the kingdom **Protista**, composed of most of the algae and protozoa, and the last is the kingdom **Moneras** or **Prokaryotes**, which includes bacteria. The difference between protists and prokaryotes is that the cells of the former have a nucleus and those of the latter do not.

ERA CENOZOICA
60 MILLONES DE AÑOS
PERÍODO CENOZOICO
PLEISTOCENO
PLIOCENO
MIOCENO
OLIGOCENO
EOCENO
PALEOCENO
HOMBRE
MASTODONTE
EOHIPPUS
MEGATERIO
ERA MESOZOICA
120 MILLONES DE AÑOS
PERÍODO CRETÁCICO
130 MILLONES DE AÑOS
PERÍODO JURÁSICO
PERÍODO TRIÁSICO
MIXOSAURIO
TIRANOSAURIO
ERA PALEOZOICA 335 MILLONES DE AÑOS
PERÍODO PÉRMICO
210 MILLONES DE AÑOS
PERÍODO CARBONÍFERO
265 MILLONES DE AÑOS
PERÍODO DEVÓNICO
320 MILLONES DE AÑOS
EDAD DEL CARBÓN
DIMETRODON
EUSTHENOPTERON
PERÍODO SILÚRICO
360 MILLONES DE AÑOS
PERÍODO ORDOVICENSE
440 MILLONES DE AÑOS
PERÍODO CÁMBRICO
520 MILLONES DE AÑOS
EURIPTERUS
EÓN AGNOSTOZOICO
ERA PRECÁMBRICA
600 MILLONES DE AÑOS
EDAD DE LA VIDA DESCONOCIDA
4000 A 5000 MILLONES DE AÑOS
4500 MILLONES DE AÑOS
5000 MILLONES DE AÑOS
AUSENCIA DE SERES VIVIENTES
FORMACIÓN DE LA PARTE MÁS PROFUNDA DE LA CORTEZA TERRESTRE
PRINCIPIO APROXIMADO DE LA TIERRA COMO PLANETA

GEOLOGICAL TIME. THE STRATIGRAPHY

Do you know how long ago our planet was formed? A long, long time ago. Geologic time is the time that has elapsed from the formation of the Earth 4.6 billion years ago to the present day. Geologic time is measured in millions of years.

HOW DO WE KNOW THE HISTORY OF THE EARTH?

It is amazing that we can describe what has been happening on our planet for millions of years, but scientists have been able to find out, by studying rocks (their age and arrangement) and fossils. But scientists have been able to find out, by studying **rocks** (their age and arrangement) and **fossils.** Do you think the Earth has always been as we know it now? At different times in its history, our planet has not had the same appearance, nor the same type of rocks, and it has not had the same life forms.Earth rocks are arranged in layers. Generally, the deepest layers are the oldest. The shallowest, on the other hand, are the most modern.Some rocks have fossils. Fossils are the remains of animals or plants that lived a long time ago, and they provide information about the age of the rock and the living things that have existed on our planet.In addition, scientists have different methods that allow them to find out the age of a rock. Using these methods and studying fossils, we have been able to know the history of our planet. Do you want to know it?

WHAT ARE THE DIVISIONS OF GEOLOGIC TIME?

Geologists (scientists who study the Earth) divide the time that has elapsed since our planet was formed to the present day into different periods. In the same way that a person's life can be divided into several stages, the history of the Earth can also be divided into several time periods. A person is a baby, then a child, then a teenager, then an adult. The Geologists use the words **eon,** era, period and epoch to denote the different stages through which the Earth has passed.The history of our planet, from its origin to the present day, can be divided into three immense periods of time, which are called eons:

- The **Archean eon** began 3.8 billion years ago and ended 3.8 billion years ago. 2.5 billion years.
- The **Proterozoic eon** began 2.5 billion years ago and ended 570 million years ago.
- The **Phanerozoic eon** began 570 million years ago and continues to the present day. We live in the Phanerozoic eon.

People divide years into months, months into days, days into hours, and hours into minutes. Geologists also divide eons into smaller periods of time, which are called **eras.** Eras, in turn, are made up of **periods,** and each of these periods is divided into several **epochs.** We live in an epoch called the Holocene, in the Quaternary period, within the Cenozoic era, which is included in the Phanerozoic eon.

WHAT HAS HAPPENED SINCE THE EARTH WAS FORMED?

The entire period of time that elapsed before the Phanerozoic eon, that is, from the time the Earth was formed until 570 million years ago, is called the **Precambrian.** The atmosphere and the oceans were formed during this period. Small organisms also appeared and changed the Earth's atmosphere. At the end of the Precambrian period, the air was similar to the air we breathe today.

The **earliest fossils of living things** date back 3.8 billion years. But the first animals and plants that have left larger fossils date back to 570 million years ago, some 3 billion years after the appearance of life on Earth. The name phanerozoic comes precisely from the presence of these large fossils: it is formed by the Greek words phaneros, which means 'visible', and zêon, which means 'animal' or 'living being'.

By studying the different fossils found in rocks, **paleontologists** (scientists who study fossils) have divided the Phanerozoic eon into eras and periods. The Phanerozoic eon is divided into three eras, called the **Paleozoic, Mesozoic**, and **Cenozoic.** Do you know what happened during that time?

LIVING BEINGS CONQUER THE TERRESTRIAL ENVIRONMENT

The **Paleozoic Era** began 570 million years ago and ended about 240 million years ago. During that period of time many living things inhabited our planet. Numerous types of invertebrates and marine algae filled the oceans. Plants invaded the terrestrial environment about 475 million years ago. While new plant species appeared on land, numerous fish swam in the seas.Animals also came out of the sea and settled on land. The first animals that conquered the terrestrial environment were similar to today's scorpions. Terrestrial insects also appeared. Amphibians, the first air-breathing vertebrates, evolved from fish. Later, reptiles emerged. Some 320 million years ago, the earth was covered with lush vegetation consisting of ferns and trees; giant dragonflies flew in the sky, and lizard-like creatures scurried along the ground.However, despite all this explosion of life, many organisms disappeared at the end of the Paleozoic era. A wave of extinctions - the largest in the planet's history - put an end to large groups of animals and plants. Scientists are trying to figure out what happened.

THE AGE OF DINOSAURS

The **Mesozoic Era** began about 240 million years ago and ended 65 million years ago. **Dinosaurs** lived during this era, which is divided into three periods, called the **Triassic, Jurassic** and **Cretaceous.** The movie Jurassic Park is named after this period. Larger dinosaurs, such as Apatosaurus, lived in the Jurassic and Cretaceous periods. The first mammals and birds, as well as the first flowering plants, also appeared during the Mesozoic. However, the end of the Mesozoic era also saw numerous extinctions. All the dinosaurs disappeared. Scientists think that the impact of a large space rock on Earth caused changes on our planet, which led to the disappearance of these reptiles. Mammals, on the other hand, survived.

AFTER THE DINOSAURS

The **Cenozoic Era** began about 65 million years ago and continues to the present day. Numerous types of mammals appeared during this era. As time went on, larger species emerged. However, some of the mammals and birds that lived about 55 million years ago were very different from those of today. There was a species of giant flightless parrots, as tall as you are! Geologists divide the Cenozoic era into two periods, called the **Tertiary** and **Quaternary**. Horses, rhinoceroses, cats, dogs and giant mammoths appeared during the Tertiary period.

WHEN DID HUMAN BEINGS APPEAR?

The first ancestors of human beings appeared about 6 million years ago, during the Tertiary period. They were smaller, hairier and less intelligent than today's humans.

The next period, the Quaternary, is further divided into two epochs, the **Pleistocene** and the **Holocene.** Present-day humans did not appear until about 120,000 years ago, in the Pleistocene. That is not a long time, if we compare it with life on Earth. If we were to imagine that the entire life of our planet has elapsed in 24 hours, human beings would have been able to live in the Pleistocene.they would only be present for the last three seconds!During the Pleistocene, huge sheets of ice covered the Earth. These periods of time are called **glacial periods.** The last glacial period ended about 10,000 years ago, when the Holocene began. We live in the Holocene; it is not over yet.

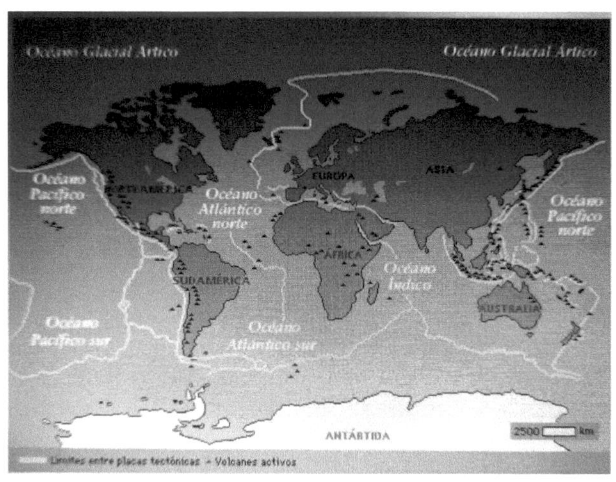
Océano Glacial Ártico
Océano Glacial Ártico
Océano Pacífico norte
Océano Atlántico norte
EUROPA
ASIA
Océano Pacífico norte
ÁFRICA
SUDAMÉRICA
Océano Índico
AUSTRALIA
Océano Pacífico sur
Océano Atlántico sur
ANTÁRTIDA
2500 km
Límites entre placas tectónicas ▲ Volcanes activos

PLATE TECTONICS

You've probably done a puzzle before. You will have noticed that some pieces fit perfectly into others. Well, although it may surprise you, something similar happens with some continents; for example, with Africa and South America. If you look at a map, you will see that the eastern edge of Brazil fits quite well into the Gulf of Guinea.

CONTINENTS MOVE

Did you know that millions of years ago on Earth there was only one large continent called **Pangaea**? Then, as time went by, the continents separated and oceans appeared between them. In the future, the continents will continue to move, so the map of the world as we know it will change. This is what the theory of the **drift of the continents,** enunciated at the beginning of the 20th century by **Alfred Wegener**, says.One proof of this theory is that the coasts of Africa and South America fit into each other, as in a puzzle. Another proof is the existence of **fossils** of the same species and common rock formations on different continents.

THE PLATES OF THE EARTH'S CRUST

Have you seen how water boils in a pot? The water moves continuously, rising and falling. If we were to place a piece of cork in the water, it would move endlessly. Well, something similar happens with the rocky materials inside the Earth.The Earth's surface is made up of a series of 'puzzle pieces', called **tectonic or lithospheric plates,** which fit together. These plates are large blocks of rock. The plates under the seafloor are called **oceanic plates,** and those under the surface of the continents are called **continental plates.** Underneath the plates is magma (rocks melted by the very high temperatures), which escapes to the surface through fissures. To explain the drift of continents or the appearance of mountain ranges there is a theory called **plate tectonic theory**. According to this theory:

- Beneath the Earth's surface are large blocks or rocky plates.
- The plates move slowly. Therefore, some of them collide with others, others

others separate, others slip away...

- Who moves these plates? The 'engine' that moves the plates is the Earth's internal heat. Under the plates, some molten rocks rise and fall, similar to the movement of boiling water in a container.
- At plate boundaries there are volcanoes, earth tremors and mountain ranges are formed.

WHEN THE PLATES MOVE

What happens when two freight trains collide? The bodies of both are deformed, there are 'folds' in the wagons, some parts break, and so on. Something similar happens on Earth when two plates collide. Let's see what happens when the plates move and how the landscape changes:

- **Formation of mountain ranges.** The collision of two continental plates forms mountain ranges. When the plates collide, the terrain rises, forming rows of mountains. Deformations of the earth's surface also occur: **folding** of the terrain and ruptures **(faults).**
- **Formation of oceans.** The separation of two continental plates leads to the formation of oceans. In this case, a rift called an **oceanic ridge** can form between the two plates, through which molten material (magma) from the Earth's interior flows out. This material forms new rocks. At the bottom of the Atlantic Ocean, for example, there is an oceanic ridge through which magma (molten material) continuously flows. This is why the coasts of South America and Africa are moving apart. The drifting away of the plates causes the continents to drift.
- **Formation of oceanic trenches and volcanoes.** The collision of an oceanic plate and a continental plate gives rise to oceanic trenches and volcanoes. In this case, the oceanic plate sinks under the continental plate. This produces earthquakes and numerous volcanoes. In onshore volcanoes, molten material from the Earth's interior is brought to the surface. Underwater volcanoes also expel gases and rocks from the Earth's interior and, in some cases, can form islands. In 1928, in the Indian Ocean, an underwater volcano produced the appearance of a new island: Anak Krakatoa.
- **Earth tremors.** The lateral displacement of two continental plates causes numerous earth tremors. This occurs, for example, on the west coast of the United States, on the San Andreas fault, where there are frequent earth tremors.

THE ROCKS

The statue of the Venus de Milo, Michelangelo's statue of David and the Taj Mahal building in India have something in common. The material from which they were built is the same: a rock called marble. This material is natural: it is "manufactured" by nature.

ROCKS SURROUND US ON ALL SIDES

When you walk down the street, look at the objects around you. You will see many rocks. If you look at the ground you will see different types of sand grains and rounded stones of different sizes. Bricks in buildings (clay), statues in the street (marble or granite), and the roof of your school (slate) are also made of rocks.A rock is a substance composed of different **minerals** found in nature. For example, granite is a very abundant rock on Earth. It is made up of three minerals: quartz, feldspar and mica. But other minerals can also appear and give granite other colors.

TYPES OF ROCKS

Look closely at a photo of granite. Now remember where you have seen this material: on floors, in statues, fountains... This material is widely used because it is hard and very abundant. But do you know where granite is formed? Inside the Earth!
As there is a great variety of rocks, it is best to classify them in order to study them. Rocks are classified into three main groups: magmatic rocks, sedimentary rocks and metamorphic rocks.

- **Magmatic or igneous rocks.** They are formed when magma expelled to the exterior of the Earth by **volcanoes** cools. On the Earth's surface this magma cools and solidifies, forming rocks such as basalt. These rocks are also called **effusive igneous rocks**. Magma can also cool in the interior of the Earth's crust. Earth forming rocks such as granite. These rocks are called igneous intrusives.
- **Sedimentary rocks.** From the rocks on the earth's surface, small pieces of different sizes are detached due to the erosive action of wind or water. These small pieces of rock are transported by water, wind and ice

to other places, where they are deposited, i.e., sediment. Over time, these sediments are compressed, forming sedimentary rocks, such as limestone. In these rocks there are different layers or strata. Approximately 75% of the Earth is covered by sedimentary rocks.

• **Metamorphic rocks.** They are formed when other rocks (igneous, sedimentary or other metamorphic rocks) undergo variations in **pressure** or **temperature**, which causes them to change their shape, composition and structure, transforming them into a different rock, a metamorphic rock. For example, when magma surrounds a rock, it heats up and can transform into another rock, a metamorphic rock. And in the formation of a mountain, due to the movements of the **lithospheric plates,** the rocks suffer great pressures that can transform them into other metamorphic rocks. Metamorphic rocks are gneiss, which is formed from an igneous rock, granite; and schist, which is formed from another metamorphic rock, slate.

THE ROCK CYCLE

Take a rock around you and look at it carefully. Do you think anything has changed in it as you look at it? No. But other rocks do change. In fact, rocks are completing a cycle that can be summed up as follows at the following steps:

1. When hot magma flows out of the Earth through a volcano, it cools and solidifies, giving rise to **igneous** rocks.

2. Wind and water erosion separates small pieces of these igneous rocks, which are then transported and deposited in other places, such as lakes and seas, forming sediments that are compacted. **Sedimentary** rocks are thus formed.

3. When these sedimentary rocks are found at great depth, magma covers them, raising their temperature a lot, and the rocks are transformed in **metamorphic rocks**. Metamorphic rocks are also created in the process of formation from the formation process.

4. Some metamorphic rocks are buried to places where the temperature and pressure increase greatly. They can then melt and transform back into igneous rocks. This closes the cycle.

VERY IMPORTANT ROCKS: COAL AND OIL

Have you recently ridden in a car or a bus? Most of the vehicles you normally use run on fuels we call gasoline or diesel. But do you know where they come from? From **petroleum**.

Coal and oil are rocks that come from animal and vegetable remains accumulated millions of years ago. We use them for heating, in vehicles,... As you can see, they are very important in our lives, since we use them as **fuel**, i.e., they are materials that give off energy (heat) when they burn.

THE FOSSIL

Did you know that there are footprints that have been preserved for millions and millions of years without being erased? Some of them, belonging to dinosaurs, have been found in La Rioja (Spain). There are also footprints of human ancestors in Africa, which have been preserved because they were surprised by a lava flow.

FOSSILS: TRACES OF LIFE IN ROCKS

Surely you have heard of animals such as **dinosaurs** or **mammoths**. However, there are no dinosaurs or mammoths alive. So how do we know they existed, and how do we know their size, what they ate or how they lived? We have found out thanks to fossils, which are the remains of living beings that appear petrified in a rock. Bones or footprints belonging to living beings that are now extinct are also considered fossils, as are the remains of animals preserved in amber or ice. Even fossilized dinosaur bones have been discovered. The science that analyzes fossil remains is called **Paleontology.**

WHAT DO THE FOSSILS TELL US?

Fossils provide us with a lot of information:

- **Fossils tell us about the history of the Earth.** They help us to know the age of a rock by analyzing to which animal or plant the fossil corresponds. They also allow us to know how the continents were distributed on our planet millions of years ago, since we have found fossils of the same species on different continents.
- **Fossils tell us about living beings that have inhabited the Earth.** At present, there are no living mammoths or dinosaurs. However, we know that they inhabited the Earth because we have found remains of bones or footprints in rocks. Fossils tell us about living things that have existed on our planet. Many of them are very old; fossils of animals called **trilobites** have been found, which lived 500 million years ago. By studying fossils, it is possible to know in which epoch a living being lived, its size or how it moved.Fossils are the greatest source of information about living beings that have existed on Earth and have become extinct.

Thanks to them we have been able to reconstruct, for example, the skeleton of many types of dinosaurs or other animals.
In addition, fossils tell us how animals and plants have evolved. For example, fossils have been found that indicate that today's birds have evolved from some types of dinosaurs!
Fossils have also been found that belong to the ancestors of human beings **(hominids).** Thus, it has been possible to reconstruct what humans looked like thousands or millions of years ago. We know, for example, that the capacity of the skull has been increasing, which has allowed for a more developed brain and greater intelligence. Human fossils also tell us how our ancestors moved and how our species evolved to its present size and behavior. The oldest hominid fossils have been found in Africa, and are between 6 and 2 million years old.

HOW IS A FOSSIL FORMED?

If you press a block of plasticine with a coin, its drawing will be engraved. Then, by looking at the image that has been marked on the plasticine, we can identify the coin used. Something similar happens when many fossils are formed. In this case, the shape of the bones or the shell of an animal, or the trunk of a plant, is engraved in the rock.
Fossils are formed when the remains of an animal or plant are buried. They always appear in **sedimentary rocks.** Sediments bury the remains of the animal or plant, which become rock. Then, due to erosion and movement of the rocks, these remains can emerge to the surface, where there is a chance that they will be discovered by scientists. Because remains are buried much earlier on the seafloor than on land, marine animal fossils are much more abundant than land animal fossils. However, not all animals leave fossils. Normally, only the hard parts are fossilized: the skeleton, the shell, etc. Therefore, there are many extinct animals and plants about which we will never know anything. But, although there are fossils distributed all over the planet, finding them is not easy. They often appear buried, and the task of identifying and analyzing them is a delicate one. Moreover, it is very difficult to reconstruct the skeleton of an animal from the remains of bones scattered on the ground. However, since fossils teach us a lot about the life of extinct animals and plants, it is worth the effort.

LIVING BEINGS OF THE PAST

In 1977 scientists found in Siberia (Russia) the frozen remains of an animal very similar to an elephant. It was a mammoth calf. But this discovery was unique, because the mammoth is an extinct animal.MThe last mammoths lived on Earth 10,000 years ago! SPECIES THAT EMERGE AND BECOME EXTINCT Do you know when dinosaurs lived? **Dinosaurs** were animals that lived a long time ago on Earth. There were dinosaurs of many kinds and sizes: some were herbivores, others carnivores..., but all of them disappeared 65 million years ago. Throughout the history of the Earth many other animals and plants have lived and have become extinct.

- When living beings are unable to adapt to change, they reproduce at a much slower rate than they die. Thus, fewer and fewer animals of that species remain, until they finally disappear.
- At other times, animals and plants do adapt to changes, but they undergo important variations that transform them into different animals and plants. This is how the **evolution** of living beings takes place.

Psilophytes, for example, are plants that lived on Earth during the Paleozoic (period beginning 570 million years ago and ending 245 million years ago). During the Carboniferous period (period that began 362 million years ago and ended 290 million years ago) there were large forests of ferns. The plants that lived during this period have formed the coal reserves that we have used so much as fuel.

WHY DO EXTINCTIONS OCCUR?

The giant panda is an **endangered** animal. It lives in China, in bamboo forests. As many of these forests are disappearing due to human presence, there are fewer and fewer panda bears. If things do not change much, in a few years there will be no bears left. panda on the planet. In 2002, there were only about 1,500 giant panda bears in all of China. The example of the giant panda will help us to know why the animals that we identify in many **fossils** are no longer living. Although it is difficult to know for sure, we can point to some probable causes, such as changes in climate, natural disasters, predation by other species, or activities carried out by humans.

- **Changes in the climate or relief of the environment.** For example, in long periods of drought, vegetation is less abundant, and then there is a lack of food for the animals that feed on the plants. On our planet there have also been **glacial periods**, in which the temperature has decreased significantly and ice and snow have covered a large part of the planet. Animals that did not adapt to this change disappeared. Mammoths, animals with large tusks similar to today's elephants, became extinct a few years ago.10,000 years. The causes of their extinction were probably climatic changes. And also hunting by humans. The people of that time hunted mammoths to eat their meat, and to make warm clothing from their skins.
- **Natural disasters.** For example volcanic eruptions or meteorite or asteroid impacts. The extinction of the dinosaurs, about 65 million years ago, was probably due to the impact of a meteorite 10 km in diameter that completely changed the climate. The impact probably covered the sky with dust and ashes, which caused a long winter that made many plants and animals disappear.
- **Predation of other species.** A flightless bird called the **dodo**, which inhabited the island of Mauritius (Indian Ocean), became extinct in less than a century, mainly due to domestic animals brought to the island by colonists. The last dodo was seen in 1681.
- **Human aggression.** Unfortunately, the presence of humans has caused and is causing the extinction of many species. Uncontrolled hunting has wiped out some species. There are currently thousands of species in danger of extinction. In many cases, international cooperation will be necessary to prevent their definitive disappearance. The **Iberian lynx**, for example, is an animal that is exclusive to the Iberian Peninsula that inhabits the south of Spain. But it is in danger of extinction. The main cause: the human presence in its environment. There are few specimens left (approximately 200) and it is possible that in a few decades this species will have completely disappeared.

THE GREAT EXTINCTIONS

Sometimes, an animal or plant species slowly disappears from the planet. At other times, however, in a short period of time, a large number of species disappear. Throughout Earth's history, there have been several such mass extinctions.

- **The Ordovician extinction.** It happened about 440 million years ago. In it 85% of the species disappeared. Many species of **trilobites** became extinct at this time. Trilobites were marine animals, they breathed through gills and had varied sizes; although most of them were smaller than the fist of your hand.
- **The Devonian extinction.** It happened about 350 million years ago. More than 80% of the species disappeared, including many corals and trilobites.
- **The Permian extinction.** It happened about 245 million years ago. More than 90% of marine species disappeared. The last trilobites became extinct at this time.
- **The Cretaceous extinction.** It occurred about 65 million years ago. Seventy-five percent of the species disappeared. In this extinction the dinosaurs disappeared and also the **ammonites**, marine animals similar to today's squids with a spiral-shaped, coiled carapace. They lived in shallow seas, mostly near the continents.

There is evidence that large extinctions occur on our planet approximately every 26 million years. That is why many scientists think that extinctions are related to the movement of the Solar System through our galaxy, the Milky Way; or by changes in the Earth's orbit that modify the climate significantly.

THE DINOSAURS

Do you know what is the heaviest land animal that has ever existed? It is a gigantic herbivorous dinosaur that was 24 meters long, more than 12 meters high and weighed 80 tons! Its name was Brachiosaurius, and it lived 150 million years ago.

WHAT IS A DINOSAUR?

Dinosaurs are land **reptiles** that lived many millions of years ago. We know they existed because we have found **fossils** of their skeletons. There were dinosaurs of many kinds: huge, small, with scales, with large jaws, without teeth, herbivorous, carnivorous (with cutting teeth and claws to catch their prey), with wings, without wings, etc. Some relied on their four limbs to walk, while others relied only on their two hind legs. Did you know that fossil dinosaur eggs have been found? Despite their large size, many dinosaurs laid eggs! Fossilized dinosaur footprints have also been found. La Rioja, in Spain, is a region where we can find some fossilized dinosaur footprints.

IN WHAT ERA DID THEY LIVE?

The first dinosaurs appeared about 230 million years ago, in an epoch known as the **Triassic**, and evolved during the 150 million years they inhabited the Earth. At the end of the **Cretaceous**, 65 million years ago, dinosaurs disappeared from the Earth; they became extinct. The first dinosaurs were small but, little by little, larger dinosaurs appeared. They became the dominant animals. During the Jurassic, more than 150 million years ago, there were large dinosaurs, such as Diplodocus, which measured 27 meters.

FIRST DISCOVERIES

The first dinosaur described and scientifically named was **Megalosaurus**. It was a carnivorous dinosaur whose remains were found in Great Britain in 1824. At that time the remains of another huge reptile were discovered near London. Its discoverer named it **Iguanodon**, because its jaw resembled that of iguanas.

TWO LARGE GROUPS OF DINOSAURS

Later, more fossils of many different dinosaurs were found. This made it possible to reconstruct almost complete skeletons. Paleontologists have classified dinosaurs into two major groups:

Saurischians. They had pelvises similar to those of reptiles. There are two groups of saurischians.

- The theropods. They were carnivores and walked on two legs. Their forelimbs were much smaller than their hind limbs. They had crests, cutting teeth and powerful claws. Some were very small, like the cat-sized **Compsognathus.** And others were huge, like **Tyrannosaurus**, which was taller than a house!
- The sauropodomorphs. They were herbivores and walked on all four legs. You may have seen a photograph or a replica of the skeleton of a huge dinosaur with a small head and a very long neck and tail. It is a gigantic herbivorous dinosaur called **Diplodocus**.

Ornithischians. They had bird-like pelvises. They were herbivores. In the Triassic, more than 200 million years ago, there appeared small, bipedal, running and agile ornithischians. And in the Cretaceous, more than 65 million years ago, lived the Iguanodon. Another dinosaur of this class was the Triceratops, which walked on four legs. In the last 10 years, dinosaur fossils have been discovered in countries such as Spain, China and Argentina whose skeletons resemble those of birds, and some fossils even show the remains of feathers! This has confirmed a theory held by many scientists: today's birds are descended from dinosaurs!

HOW DID THE DINOSAURS BECOME EXTINCT?

65 million years ago, all dinosaurs abruptly disappeared.

How? It is not known with certainty. Scientists believe that the cause was the impact of a huge **meteorite** (several kilometers in diameter) against the Earth. One piece of evidence supporting the meteorite theory is the existence of a crater of about 200 km in diameter, found in the Yucatan Peninsula in Mexico. But don't think that the meteorite hit all the dinosaurs. At that time, there were dinosaurs all over the planet. But the impact caused a cloud of dust that darkened the sky for months, causing the death of many marine and terrestrial plants. Without plants, many herbivorous animals ran out of food and died. And then carnivores also ran out of food.... So it was not only the dinosaurs that became extinct. Many other animals also disappeared, such as the **ammonites**. We know that all these animals existed because fossils have been found.

THE ORIGIN OF SPECIES HUMAN

Did you know that until just two hundred years ago almost everyone believed that human beings appeared on Earth as we are today? It was Charles Darwin who, in his book The Origin of Species (1859), put forward the theory of evolution, that is, the idea that all living things developed by evolving through natural selection. Although he was criticized using a phrase he never wrote: "man descended from the ape", what he actually said is that human beings share many characteristics with apes and that both (humans and apes) descended from a common ancestor.

WHAT DO HUMANS AND MONKEYS HAVE IN COMMON?

Between 10 and 5 million years ago, an ancestor common to man and the higher apes lived on Earth. Therefore, at some point in that period, there was a separation between the **hominid** line that leads to us and the ape line that leads to today's apes. This coincided with a change in the Earth's climate, which resulted in colder and drier conditions, reducing the African forests and creating large areas of savannah or clear forest.

THE FIRST HOMINIDS

Between 6 and 2 million years ago, hominids appeared in East Africa, the **australopithecines,** small in stature and brain, but with two novel features: they walked on two legs and had small fangs. Another group appeared 2.5 million years ago, the **Homo habilis**, so called because its members were the first to be able to make stone tools, with which they could skin large dead animals (scavenging) or hunt small animals. With a slightly larger brain and greater stature than the previous group, this was the first representative of the genus Homo.Around 1.8 million years ago, and as an evolution of Homo habilis, **Homo erectus** emerged, with a larger brain (1,000 cm^3) and with the ability to build new stone tools, called **bifaces** (axes and axes). hand). He was the first to learn to light and use fire. The groups of Homo erectus, following animal migrations, left Africa for the first time and dispersed throughout Asia and Europe. Those established in Europe, with some of their own traits, are known as

Homo heidelbergensis.

NEANDERTHAL MAN AND MODERN HUMANS

Between 200,000 and 150,000 years ago, **Neanderthal man** (Homo sapiens neanderthalensis) appeared in Europe and the Near East, as an evolution of the later heidelbergensis. Neanderthals were short, very robust and had large brains (1,500 cm^3), even larger than ours. They improved stone tools (Mousterian industries); they lived in caves and open-air camps, and were the first to bury their dead.

About 120,000 years ago, the first modern humans, **Homo sapiens sapiens**, emerged, again in Africa. They were tall; dark-skinned, adapted to the tropics, and practically the same as us; in fact, we are the same species! They made more complex stone and bone tools; they invented the first ornaments (necklaces and pendants), and created the first artistic manifestations (engravings and paintings).

Modern humans, our direct ancestors, came out of Africa ('second out of Africa') and spread across Asia replacing the Homo erectus populations. They were the first settlers of Australia and the first navigators, around 60,000 years ago. They arrived in America from the tip of Siberia, crossing the ice pass that joined Asia and America, between 30,000 and 15,000 years ago, and in a few millennia they occupied the entire continent.

About 40,000 years ago, the first modern humans (also called **Cro-Magnon** men) entered Europe and 'collided' with the Neanderthals, who, for various reasons, became extinct, after a few millennia of 'contact', just over 30,000 years ago. Thus, only one single human species remained on Earth: us! Modern humans developed the great parietal art (the one depicted on walls and ceilings) in many European and some Asian caves.

THE PREHISTORY

You may have seen movies showing men fighting dinosaurs. In reality that could not have happened, because dinosaurs disappeared from the Earth long before any human beings existed. Humans began to emerge at the beginning of a long period known as prehistory.

WHAT IS PREHISTORY?

Prehistory is a period in the history of mankind. The first and the longest lasting: from the beginning of the process of **human evolution** until the appearance of written texts. We could say, therefore, that a people lives in prehistory until it writes.

Archaeology deals with the study of prehistory. Archaeologists search for and analyze the material remains left by the first human beings. Thanks to archaeology, we know something of the life of our ancestors: who they were, where they lived, what were the first instruments they made and their first works of art... As it was a very long period, prehistory is usually divided into two major stages or ages: the Stone Age and the Age of Metals.

THE STONE AGE

The Stone Age is so called because in that remote time the ancestors of human beings began to make, mainly with **stone,** their first instruments or tools. They also used other materials, such as wood, and the bones, horns and tendons of the animals they hunted.During the Stone Age, the long process of human evolution took place. When the Stone Age ended, the present-day human being already existed: scientists call us **Homo sapiens sapiens**! Along the way, many human-like species remained, which became extinct or evolved. The Stone Age is divided into three periods.

- **The Paleolithic.** During the Paleolithic, which began 2.5 million years ago, man learned to carve **stone.** Paleolithic humans were **hunter-gatherers.** What does this mean? Quite simply, they lived by hunting and fishing, and by what they gathered from plants (roots, fruits).
- **The Mesolithic.** It lasted from the end of the Paleolithic to the beginning of the Neolithic. Humans continued hunting and gathering for

subsistence.

• **The Neolithic.** Man was already polishing stone (i.e., he worked it with greater precision and could build more sophisticated tools). But the great change that took place during the Neolithic (you may hear the expression **Neolithic revolution**) is that **agriculture** was born (approximately 11,000 years ago, in 9000 BC). Because of this, human beings ceased to be nomadic (until that time, they had no fixed home) and became sedentary (thus the first **settlements** appeared). The manufacture of **ceramic** pieces also became common.

THE AGE OF METALS

There came a time in prehistory when people stopped making their tools with stone and started making them with **metals** (because they learned the necessary techniques to do so). The metal age is divided into three periods.

• **The Copper Age.** Copper was the first metal used by humans.

• **The Bronze Age.** It is so called because man began to use bronze, after learning to alloy (mix) copper with another metal: tin. The ancient cultures of Mesopotamia were born and developed during the Bronze Age. At the end of that period, the first civilizations of Greece also emerged.

• **The Iron Age.** Utensils began to be made of iron. It began in the Near East about 3,200 years ago (although in most of Europe it did not begin until about 1,300 years ago, and in America iron was not worked until the arrival of Europeans at the end of the 15th century A.D.). The Europeans of the Iron Age belonged, for the most part, to the **Celtic culture.** The Iron Age The iron age ended in almost all of Europe as the Roman conquest took place in each area.

PREHISTORIC ART

The first artistic manifestations of human beings were born during prehistoric times. Specifically, they arose in the Paleolithic period. Have you ever heard of the Paleolithic paintings in the cave of Altamira, in Spain, or the Neolithic megaliths of Stonehenge, in England? They are samples of prehistoric art, as are the bronze utensils from the archaeological site of El Argar (in the Spanish province of Almeria), or the sculptures of the Toros de Guisando (found in the Spanish province of Avila and sculpted in the Iron Age).

MOUNT GREEN

Do you want to know where the oldest evidence of human presence in the Americas has been found?

WHAT IS MONTE VERDE?

Monte Verde is a South American prehistoric archaeological site. It is located on the banks of the Chinchihuapi stream, about 35 kilometers from Puerto Montt, in southern **Chile. It** was discovered in 1976 and, since then, it has become one of the most important discoveries to know since when human beings inhabited America.

GREEN FOREST STRATA

Do you know what a geological stratum is? It is a layer of soil that contains what was buried at a particular time. In Monte Verde two different strata have been found and identified: Monte Verde II and Monte Verde I.

Monte Verde II. It is the most recent stratum (it is located on top of the other one). In Monte Verde II, an encampment of a group of hunters and gatherers who stayed there most of the year has been found. It is believed to be about 14,500 or even 15,000 years old (it would belong, therefore, to the last years of a period of prehistory called the **Upper Paleolithic**). Why is Monte Verde II important? Because it is one of the first signs of **sedentary life** in South America. When excavating, archaeologists found carved stones (the primitive tools of the local inhabitants) and some wooden and bone tools. They also found remains of huts that were built with wood.

Monte Verde I. It is the oldest stratum. In it, 26 boulders have been found that were carved by man more than 33,000 years ago (at the beginning of the Upper Paleolithic). What does this mean? It means that humans were already living in the Americas at that time. Until the discovery of Monte Verde I, it was thought that humans had arrived in America about 20,000 years ago. 22,000 years ago, when Siberian peoples crossed the Bering Strait, taking advantage of the fact that it was frozen.

BIBLIOGRAPHIC REFERENCES

Asimov, I. 1983. The terrible lizards and other scientific essays. Third Edition. Alianza Editorial. Madrid, 188 pp.

Asimov, I. 1996. The terrible lizards. Alianza Editorial, S. A. Madrid, 63 pp.

Cabrera, A. 1962. Prehistoric animals. Ed. Espasa-Calpe, S. A., Madrid. 94 pp.

Fernández, L. H. 2010. Prehistoric animals 1 and 2. Wonders of Zoology. Editorial Academia. Havana, 70 pp.

Guarch, J. J. 2002. Los Dinosaurios. Ediciones Holguín. Holguín, 80 pp.

Hernández-Muñoz, A. 2005. Expedition to the depth of the millennia. Editorial Gente Nueva. Havana, 69 pp.

Howard, J. 1981. I can read about dinosaurs. Troll Associates. Mahwah, N.J., 43 pp.

Iturralde- Vinent, M. 1989. Sal si puedes. Editorial Gente Nueva. Havana, 99 pp.

Laza, J. H. 1984. Field techniques in Vertebrate Paleontology. Com. Mus. Prov. Sci. Nat. "Florentino Ameghino". Santa Fe, Vol. 1(2): 1-71.

Ocean Institute Gallach Institute.2000. Paleontology, Vol.12. NATURAL HISTORY. Océano Grupo Editorial, S. A. Barcelona, 2251 p.

Scott, J. 1975. Introduction to Paleontology. Ed. Paraninfo. Madrid, 191 pp. Tonni, E. 1994. THE HISTORY OF A STREAM. An encounter with fossils.An awakening of science. Editorial Lumen. Buenos Aires, 23 pp.

Printed by Books on Demand GmbH, Norderstedt / Germany